SUSTAINABLE RURAL WASTEWATER MANAGEMENT IN THE PEOPLE'S REPUBLIC OF CHINA
INSTITUTIONAL, REGULATORY, AND FINANCIAL FRAMEWORKS AND STAKEHOLDER PARTICIPATION

NOVEMBER 2023

ASIAN DEVELOPMENT BANK

CONTENTS

TABLES AND FIGURES

ACKNOWLEDGMENTS

The authors of this publication—Mingyuan Fan, principal water resources specialist of the Agriculture, Food, Nature, and Rural Development Sector Office (SG-AFNR), Asian Development Bank (ADB); and Mats Andersson, ADB consultant—would like to express their great appreciation to ADB's EARD management for their valuable guidance, supervision, and support, most especially to M. Teresa Kho, director general; and Thomas Panella, Director, SG-AFNR.

For his efforts and insightful comments as peer reviewer, the authors are grateful to Jude Kohlhase, unit head, project administration, Water and Urban Development Sector Office.

This publication is based on the consultants' final report, *Institutional and Regulatory Framework for Rural Wastewater Management in the PRC* (People's Republic of China), under the Rural Vitalization—Rural Wastewater Treatment and Environmental Management Project (TA 9825-PRC), authored by Li Shuang, Yu Wencan, Joey Shen and Zhao Qihong. Fei Yue has provided guidance to the team and reviewed the final report. The expert panel of peer reviewers comprised Zhang Xiaowen, Shi Yi, Wu Yubing, Xin Zhiwei, and Ma Zhong.

Joy Quitazol-Gonzalez facilitated the publishing of this knowledge product, including the final review of the draft report. The authors would also like to thank Sophia Castillo-Plaza, EAOD; Christian Fischer, and Jesusa Quinto, SG-AFNR for their assistance.

ABBREVIATIONS

CAO	Central Agricultural Office
CHAST	child hygiene and sanitation training
CHC	community health clubs
CLTS	community-led total sanitation
CPC	Communist Party of China
MARA	Ministry of Agriculture and Rural Affairs
MEE	Ministry of Ecology and Environment
MHURD	Ministry of Housing and Urban–Rural Development
MOF	Ministry of Finance
NDRC	National Development and Reform Commission
O&M	operation and maintenance
PHAST	participatory hygiene and sanitation transformation
PRC	People's Republic of China
RWWM	rural wastewater management
SDG	Sustainable Development Goals
SLTS	school-led total sanitation
UNICEF	United Nations Children's Fund
WHO	World Health Organization

CURRENCY EQUIVALENTS

As of October 2023

CNY 1.00	=	$ 0.136
$1.00	=	CNY 7.31

EXECUTIVE SUMMARY

The purpose of this technical paper is to provide an outline of the current institutional and regulatory framework for rural wastewater management (RWWM) in the People's Republic of China (PRC). In this context, RWWM refers broadly to the provision of wastewater services (from collection to treatment and reuse of wastewater) in conjunction with the rational and sustainable use of natural resources. This paper highlights the existing institutional arrangements and recent policy developments, including regulations, funding arrangements, and education and participatory approaches. Technical wastewater management solutions and operational guidelines, including operation and maintenance (O&M) matters, will not be discussed in this paper. A summary of provincial case studies in the PRC is included in Appendix 1. The cases show the challenges faced and the achievements accomplished under different local circumstances across the country. An overview of rural sanitation approaches is outlined in Appendix 2. Rural sanitation approaches aim to improve the sanitation, hygiene, and water conditions of rural areas through proper RWWM, which are critical for the sustainable growth of rural communities. International experiences in the planning and design of rural sanitation programs are likewise discussed, including the cost of rural sanitation programs (Appendix 3).

Institutional Framework

An established decision-making and coordination mechanism for promoting RWWM exists under the direct leadership and guidance of the Communist Party of China (CPC) and the government throughout the administrative hierarchy (provincial, city, county, township, and village levels). Strategies are set at the central level, and general coordination is arranged through a division of labor among the line ministries. The implementation is accomplished at the local level, mainly through efforts at the city and county levels. Their capacity is constrained, however, by local conditions, such as the availability of natural resources, their development status, and funding.

At the central level, the Central Agricultural Office (CAO) is the foremost body for policymaking and coordination of issues relating to the livelihood of farmers, agricultural development, and rural development broadly. The CAO has an office in the Ministry of Agriculture and Rural Affairs (MARA) which is in charge of daily operations and is headed by the agriculture minister.

The Guidance on Promoting Rural Wastewater Management, jointly issued in 2019 by the CAO and eight concerned ministries, sets out the three-level governance structure for improving RWWM. Strategic targets are set at the central level, delivery responsibility is assigned to the provincial level, and implementation is the responsibility of the cities and counties. MARA leads the improvement of rural living conditions, while the Ministry of Ecology and Environment (MEE) is responsible for improving the RWWM.

Policy and Regulatory Framework

Since 2016, 15 key policy documents for improved RWWM have been issued by the central government of the PRC. These are found in Appendix 4. They include strategies, programs, action plans, guidelines on technical norms and standards, discharge standards, institutional arrangements, and guidance on encouraging the private sector to participate in providing RWWM services.

There are no specific laws for rural domestic wastewater treatment at the national level in the PRC, but many provinces have issued regulations, including local discharge standards. Zhejiang Province in the country's southeast is a pioneer in RWWM; it issued the first local regulation on RWWM in January 2020, which provides a useful reference for other provinces. Appendix 5 provides a list of local policy documents, strategies, and regulations for improved RWWM issued by 16 provincial governments. They include provincial-level programs, specific action plans, operational guidelines, and monitoring and assessment mechanisms.

RWWM policies and regulations are still under development in the PRC, following a phased approach by implementing demonstration cases in some provinces before upscaling and replication. Under the government's economic development strategy (i.e., the 13th Five-Year Plan [2016–2020]), the focus was on developing national principles and standards to guide local decision-making, particularly executing demonstration cases at the county level that align with the new principles and standards. The 14th Five-Year Plan (2021–2025) is oriented toward upscaling and replication, with an emphasis on adapting to local conditions and demand. The overall aim under the current 14th plan is for communities to achieve a functional system of ensured service standards and operational sustainability across the country.

To promote sustainable rural development and improve rural livelihoods, the general offices of the CPC Central Committee and the State Council jointly released in February 2018 the Three-Year Action Plan for Rural Living Environment Renovation. The action plan includes requirements for wastewater management and toilet renovation in rural areas. The central government launched the Rural Vitalization Strategy in 2018, and then the Rural Vitalization Strategy Plan (2018–2022).

Financing

The National Development and Reform Commission (NDRC) leads the development of the pricing mechanism for rural wastewater treatment services and the exploration of appropriate cost-sharing mechanisms among stakeholders. RWWM is mainly financed at the local level, complemented by funding support from the central government and contributions from the private sector. The central government support is managed by the Ministry of Finance, in cooperation with the MEE. They provide special funds for improving rural environmental management, including support for solid waste management, rural wastewater management, water source protection, and rehabilitation.

At the provincial level, the CPC and relevant government are responsible for ensuring an enabling environment for the implementation of RWWM. The CPC and the city and county governments are the key entities for preparing project proposals and ensuring the successful implementation of RWWM. The township level organizes and supervises the day-to-day construction and operation of the rural wastewater treatment facilities. At the village level, the CPC promotes the awareness of good hygiene and rural sanitation practices among farmers.

Figure: Financial Sources from Different Sources

| Support from the central government | Support from local governments | Private sector participation | Funding from financial institutions | Payments by rural households |

Source: Asian Development Bank.

To build rural wastewater treatment facilities and improve their efficiency, local (provincial, city, and county) governments can integrate financial resources from different sources. They may issue bonds for RWWM and are encouraged to work with the private sector on the subject through public–private partnership arrangements. In-kind contributions from villagers, investments, or donations from the private sector and individuals are also encouraged. To ensure operational sustainability, some provinces require that cities or counties include an allocation for upgrading and O&M of the rural wastewater treatment facilities in their annual budgets.

Financing channels include the following:

- **Support from the central government.** The city or county governments are the primary source of financing for improved RWWM. Central government provides financing support to encourage the concerned local governments, particularly those in less developed regions, to protect and rehabilitate water sources and improve their rural solid waste and wastewater management practices.
- **Support from local governments.** Local governments are required to ensure funding for improved rural living conditions, mainly through the issuance of bonds or revenues from land development. County governments can use resources from different channels at their discretion to ensure the implementation of development programs.
- **Private sector participation.** While private sector investors or service providers are encouraged, the challenge is to create a steady stream of revenues to ensure that they can get an appropriate return on their investments over time.
- **Funding from financial institutions.** Loans from the commercial financial sector face similar challenges as private sector investments: ensuring an adequate return while being manageable for the local governments or communities as borrowers.
- **Payments by rural households.** While this is the solution in the medium and longer term, presently in most provinces, rural households shoulder a small portion of the O&M cost of treatment facilities due to affordability constraints.

In summary, RWWM at present requires funding support from the public sector for both construction and O&M of treatment facilities. Tariff payments by households are unlikely sufficient as the primary source of short-term financing. Private sector participation needs long-term financing support from the government to ensure appropriate returns on their investments.

Conclusions and Recommendations

The institutional arrangement for promoting and supporting RWWM among the central government ministries needs to be further clarified, particularly between the different layers of the government hierarchy at the central, provincial, and local levels. It is critical to improve coordination to ensure effective implementation of RWWM.

Legislation and corresponding regulations for RWWM should be formulated at the central, provincial, and local levels to provide a robust legal and regulatory framework to enable and sustain improved rural wastewater treatment and management performance. Further, a monitoring system must ensure transparency and measure performance accurately. These actions would facilitate the advancement of sustainable development of the RWWM in the country.

The most appropriate cost recovery mechanisms and business models should continue to be explored to encourage the private sector to engage in RWWM, as investors and/or service providers.

INTRODUCTION

Background

The People's Republic of China (PRC) has recorded impressive progress in reducing poverty and improving nutrition.[1] In 2021, the PRC government announced that according to its national poverty threshold, it had expunged extreme poverty.[2] Between 2015 and 2020, access in rural areas to drinking water improved from 84.5% to 89.7%, and access to basic and safely managed sanitation improved from 76.5% to 87.9%.[3] Despite such progress, many challenges remain in rural areas. Intensive agriculture, economic development in villages and small towns, and the growth of tourism have increased environmental and ecological pressures. Both point source pollution (sewage) and nonpoint source pollution (agriculture-based and other runoff) have impaired water safety in many areas. Poorly maintained or a lack of public infrastructure and sanitation services affect people's living conditions and have resulted in the direct discharge of domestic pollutants into groundwater and surface water bodies.[4]

Rural wastewater treatment is crucial to improve the PRC's rural living environment (e.g., to prevent gastrointestinal diseases, cholera, typhoid, stunting, etc).[5] It is an important measure to improve the effectiveness of the implementation of the country's rural vitalization strategy and an inherent requirement to improve the quality of the rural water environment.[6] Households that are economically disadvantaged and geographically isolated have greater difficulty obtaining access to wastewater services. In recent years, all provinces in the PRC have actively promoted rural wastewater treatment and achieved significant results. This has played a vital role in improving the ecological environment of rural areas, enhancing the quality of life of farmers, and promoting agricultural and rural modernization.

---

[1] Based on the country's official poverty line of less than $2.15 a day (2017 purchasing power parity), the country's poverty ration decreased from 72.0% in 1990 to 0.1% in 2019; World Bank. Poverty and Inequality Platform. https://data.worldbank.org/indicator/SI.POV.DDAY?locations=CN (accessed 16 May 2023).

[2] World Bank and Development Research Center of the State Council of the PRC. 2022. *Four Decades of Poverty Reduction in China: Drivers, Insights for the World, and the Way Ahead*. Washington, DC: World Bank. https://thedocs.worldbank.org/en/doc/bdadc16a4f5c1c88a839c0f905cde802-0070012022/original/Poverty-Synthesis-Report-final.pdf.

[3] World Health Organization (WHO)/United Nations Children's Fund (UNICEF) Joint Monitoring Programme. Water, Sanitation and Hygiene (WASH). Rural and Urban Sanitation Service Levels, 2015 and 2020: Household Data by Service Level. https://washdata.org/data/household#!/ (accessed 17 May 2023).

[4] It is estimated that 85% of the nonpoint source pollution in the PRC is derived from rural sources, particularly the discharge of pollutants such as total nitrogen and total phosphorus from agricultural activities. The dispersed nature of these activities makes the control of nonpoint source pollution a major challenge. However, this topic is not addressed in this document, which focuses on domestic (household) sanitation.

[5] Rural wastewater treatment refers to domestic (household) wastewater, not industrial or agricultural wastewater management.

[6] Central Committee of the Communist Party of China and the State Council of the PRC. 2018. *Guidelines for the Implementation of the Rural Revitalization Strategy*. Beijing.

The Government of the PRC identifies rural wastewater management (RWWM) as an important issue and is reflected in the following five-year plans: (i) 13th Five-Year Plan (2016–2020),[7] which focused on developing national principles and standards to guide implementation, particularly demonstration cases at the county level; and (ii) 14th Five-Year Plan (2021–2025),[8] which aims to upscale and replicate successful cases using a phased approach with greater emphasis on project designs based on local conditions and needs. The overall aim of the current plan is for communities to achieve functional systems with ensured service standards and operational sustainability across the country.

At the end of 2010, the PRC's rural population was 552 million (39% of the total population)[9] and scattered across about 500,000 administrative villages and 2.5 million natural villages.[10] The PRC's annual investment in rural wastewater treatment increased from CNY12.6 billion in 2016 to CNY75.4 billion in 2020—a sixfold increase.[11] During the same period, treatment capacity doubled from 25.3 million cubic meters per day to 51.0 million cubic meters per day. In 2006, the national average annual rate of increase in rural wastewater services was reported at 1%. It increased to 6% by 2010 and to 10% by 2014. At the end of 2018, rural wastewater service coverage ranged from 10% to 90%, depending on the region. There is still significant room for improvement, particularly in villages.

Responsibility for promoting RWWM falls under both the Communist Party of China (CPC) and the administrative hierarchy of the government (at the provincial, city, county, township, and village levels). At the national level, the government has established a comprehensive management structure for the promotion of rural wastewater treatment, with relevant departments coordinated by the CAO in the Ministry of Agriculture and Rural Affairs (MARA).[12] The CAO has issued policies and regulations to guide the country on the subject, and considerable results have been achieved in recent years.

The national government's policy on rural wastewater treatment is based on the principle of adapting measures to local conditions and progressing step by step. The 13th Five-Year Plan focused on demonstration projects, construction of facilities in stages, and principles and guidelines from the central government. The 14th Five-Year Plan builds on the 13th Five-Year Plan, addressing construction, management, and operations. It aims to establish a long-term policy system and promotion mechanism. To promote sustainable rural development and improve people's livelihoods in rural areas, the General Office of the CPC Central Committee and the General Office of the State Council jointly released the Three-Year Action Plan for Rural Living Environment Renovation in February 2018, with requirements on wastewater management and toilet renovation in rural areas.[13] The central

---

[7] Government of the PRC, Ministry of Environmental Protection, and Ministry of Finance. 2017. *The 13th Five-Year Plan (2016–2020) for Comprehensive Improvement of the National Rural Environment*. Beijing.

[8] Government of the PRC. 2021. *The 14th Five-Year Plan (2021–2025) for National Economic and Social Development and the Long-Range Objectives Through the Year 2035*. Beijing.

[9] National Bureau of Statistics of China. 2021. *2020 China Statistical Yearbook*. Beijing: China Statistics Press.

[10] An administrative village is a level 5 administrative division in the PRC, which falls under the township (level 4), county (level 3), city (level 2), and province (level 1). Some villages, however, are not administrative villages but natural villages. Natural villages are communities formed spontaneously through settlement choices over a long period of time. They are "the basic rural administrative units, the lowest-level collective land management units, and the primary locus of collective labor and income distribution of the Maoist era." (Yi Wu. 2016. Land Rights, Political Differentiation, and China's Changing Land Market: Bounded Collectivism and Contemporary Village Administration. *The Asia-Pacific Journal*. 14 (1). No. 3. Article ID 4842).

[11] Government of the PRC, Ministry of Housing and Urban–Rural Development. 2021. *China Urban–Rural Construction Statistical Yearbook 2020*. Beijing: China Statistics Press.

[12] MARA was established in 2018 by merging the functions of the previous Ministry of Agriculture with the agriculture-related responsibilities of several other agencies (i.e., the Ministry of Finance, the Ministry of Land and Resources, the Ministry of Water Resources, and the National Development and Reform Commission) to streamline the government and improve its efficiency in promoting rural development.

[13] E. Tangen. 2018. China Unveils Action Plan for Improving Rural Living Environment. *China Today*. 28 February.

government, in 2018, also launched the Rural Vitalization Strategy and issued the Rural Vitalization Strategy Plan (2018–2022).[14] The Rural Vitalization Strategy Plan promotes localized rural sanitation management and innovative RWWM models to rehabilitate ecological environments and improve the character of the community, including the livability of rural villages.

There is no specific national law nor related regulations for rural domestic wastewater in the PRC. Neither the Water Law (2002) nor the Water Pollution Prevention and Control Law (amended in 2017) cover the issue nor specify the responsible departments for rural wastewater in the country. The subject is addressed indirectly in national public health legislation (e.g., on discharge standards) and local regulations. Some provinces have issued regulations for rural domestic wastewater treatment; however, these lack legislative basis, and the management levels vary significantly on the subject. For example, Zhejiang Province has formed an advanced policy system regulating the construction, management, and operation and maintenance (O&M) of rural domestic wastewater treatment systems. Some provinces are still in the early stages of facility construction.

Technical Considerations

Rural villages produce wastewater, the volume of which is closely associated with their water usage. Increased water supply creates a need for managing a greater volume of wastewater. In the PRC, rural areas generate wastewater that is about 60%–80% of their water usage volume.

Rural wastewater services in the PRC can be classified as decentralized or centralized systems. A decentralized system is a system installed and managed on-site, at the household level or in the household. A decentralized system can also be a system where households are grouped and connected to a small neighborhood system. This is known as a mixed system. A centralized wastewater system typically comprises a community-level system for sewage collection, treatment, and disposal.

Any RWWM initiative should assess the costs and benefits of a decentralized system versus a centralized one. Some technologies can be complex and costly to operate and maintain due to their required consumable materials (e.g., chemicals) and spare parts (e.g., filters). To arrive at economic and more sustainable solutions, in addition to the community needs, the location, geography, soil type, hydrogeology, climate, and other environmental factors need to be examined as well. Technical solutions and operational guidelines of RWWM are not covered in this paper.

Provincial Cases

The case studies of RWWM in Jiangsu, Shanxi, and Zhejiang provinces are summarized in Appendix 1, including financing mechanisms applied and private sector participation. RWWM is at the most advanced development stage in Zhejiang Province, while Jiangsu Province is at the mid-range and presently more advanced than Shanxi Province. Due to historical reasons, some local governments use an institutional arrangement that is quite different from what is established at the provincial or national level. For example, the Ministry of Housing and Urban–Rural Development (MHURD) is responsible for RWWM implementation in Zhejiang Province, while MEE is the leading agency in Shanxi Province.

---

[14] Xinhua. 2018. China Releases Five-Year Plan on Rural Vitalization Strategy. *The State Council of the People's Republic of China.* 26 September.

Even in the case of Zhejiang Province, which has the most advantageous natural resources, it is difficult to enforce the "polluter-pays principle." It has instead adopted a gradual approach for its (individual and cluster) household payment system to cover part of the O&M costs for its rural wastewater treatment services.

Local governments tend to either group wastewater treatment services for clusters of villages and townships or combine the service with other public services (e.g., water supply, power, or solid waste disposal) to ensure economies of scale for private sector investors. Considerations for clustering services are dependent on the capacity of a village to manage assets and deliver the services. Some local governments, such as in Jiangsu, Shanxi, and Zhejiang provinces, have developed institutional arrangements and monitoring techniques for the supervision of private sector service providers to ensure quality service. These cases serve as demonstration cases for potential replication in other parts of the country.

Rural Sanitation Approaches

An array of approaches are employed to improve the sanitation conditions of rural communities.[15] Appendix 2 provides an overview of these. Some approaches emphasize changing individual behaviors, while others focus on collective change and generating awareness. Organizations and governments also sometimes mandate the types of approaches to be used in certain geographical areas or for a particular context. Directives for the strict adherence to these mandated approaches can inhibit adaptation, modification, or innovation, or even hamper proper documentation of "blended" and "unorthodox" approaches. As a result, the "whys" and "hows" behind the successes and failures of rural sanitation approaches can be obscured, thwarting moves toward more context-appropriate, flexible, and responsive rural sanitation programming.

Large-scale rural sanitation programs for a particular geographical area with international support have usually applied the following principles:

- using **an area-wide approach,** with emphasis on social inclusion, complete coverage, and sustainable services for the poorest people;
- striving for **access for all** (universal coverage), by addressing equity and processes of exclusion through innovation and guaranteeing gender equality;
- promoting **sustainable behavioral change,** particularly in terms of clean drinking water, sanitation, and hygiene; and
- implementing **systems change,** both in the institutions and financial systems where services are embedded, and in consideration of the checks and balances of the systems and their interconnectedness.

---

[15]   Based on the report *Review of Rural Sanitation Approaches*, published in August 2017 and commissioned by WaterAid, Plan International, and the United Nations Children's Fund (UNICEF).

Designing an area-wide rural sanitation program (e.g., for one or more counties) should aim to meet the sanitation target of the United Nation's Sustainable Development Goals (SDG 6)[16] by (i) promoting universal access to safely managed sanitation and hygiene, (ii) eliminating open defecation, (iii) progressively reducing inequalities among population subgroups, and (iv) alleviating sanitation and hygiene burdens for women and girls. This will require inclusive and effective programming, as well as attention to the scale, sustainability,[17] and equity of program design, selection of implementation approach, and monitoring of sustained outcomes.

Key Issues and Challenges

Since 2010, rural sanitation and wastewater management in the PRC has improved significantly as a result of intensive government investment. Many villages have installed sanitary toilets, which are either connected to public sewers or on-site wastewater treatment facilities. Nevertheless, practical problems such as insufficient local attention and the difficulty of changing the habits of residents persist, and regulatory developments are slow. Some of these issues are summarized as follows.

Institutional Setting

- There are ambiguities regarding the responsibilities of rural wastewater treatment. The 2018 national institutional reform places the responsibility for constructing rural wastewater treatment facilities on the MEE at the national level. The lead department, however, at the provincial level can vary—e.g., the MEE is the leading agency in Shanxi Province, while the MHURD is the responsible agency in Zhejiang Province. This may increase the difficulty of implementing policies and regulations.

Policies and Regulations

- There is no specific law on rural wastewater and sanitation in the PRC. The subject is addressed indirectly in national public health legislation (e.g., on discharge standards) and in local regulations.
- The existing policies and regulations on RWMM are not responsive and are not clear enough to support effective planning and operations.
- The rural wastewater discharge standards determine the process selection, equipment configuration, and O&M costs of treatment facilities. Although the authorities emphasize that such standards should be adapted to local conditions (including local economic development level and management capabilities) and need to have technical feasibility, most of the discharge standards applied by local government still reflect higher standards and stricter requirements than necessary. This increases investment and O&M costs and may lead to an unsustainable development situation.
- The provincial enforcement of effluent standards is weak.

---

[16] By 2030, end open defecation and achieve universal access to adequate and proper sanitation and hygiene, particularly focusing on the needs of vulnerable groups, including women and girls.

[17] L. Seghezzo. 2009. The Five Dimensions of Sustainability. *Environmental Politics*. The five dimensions of sustainability are (i) *institutional sustainability* (Do institutions fulfill their roles and responsibilities continuously over time?); (ii) *financial sustainability* (Are there adequate finances for institutions, services, and outcomes over time?); (iii) *functional sustainability* (Do facilities and services continue to function over time?); (iv) *equity sustainability* (Are services and outcomes equitable over time?); and (v) *environmental sustainability*.

Financing

- The financing approaches of rural wastewater treatment facilities are unsustainable, relying heavily on government financing and subsidies. The funding for rural wastewater treatment is mainly from local governments, with some rewards and subsidies provided by the central government.
- Due to the uncertain returns of rural wastewater projects, enthusiasm for private participation is low.
- A tariff payment system by farmers is being piloted in some areas where conditions permit, but considering the limited capacity to pay off most rural villagers in the PRC, this is not expected to be the primary source of funding in the long-term.
- A cost recovery mechanism for sufficient O&M of facilities to ensure its sustainability is lacking.

Stakeholder Participation

- There is inadequate stakeholder participation in, public awareness of, and ownership of RWWM initiatives, which hamper the delivery of good environmental and public health outcomes.

PLANNING AND DESIGN OF RURAL SANITATION PROGRAMS

Introduction

The World Health Organization (WHO)/United Nations Children's Fund (UNICEF) Joint Monitoring Programme for Water Supply, Sanitation, and Hygiene defines improved sanitation as facilities that are used and ensure hygienic separation of human excreta from human contact. These facilities range from flush or pour–flush toilets or latrines, pit latrines with a slab, or ventilated improved pit latrines, to a septic tank, composting toilet, and piped sewer system.[18] According to the program, in 2020, the PRC's sanitation facility coverage in rural areas was 90.6%, compared to 97.6% in urban areas. These rural sanitation facilities include sewers (40.4%), septic tanks (15.2%), and latrines (35.0%).[19] Table 2.1 presents the progress in sanitation in the PRC's rural and urban areas in 2015 and 2020. This shows that sanitation problems are most acute in rural areas.

Table 1: Sanitation Progress in the People's Republic of China, 2015 and 2020
(%)

Sanitation System	Rural Areas		Urban Areas	
	2015	2020	2015	2020
Open Defecation	1.2	0.3	0.2	0.3
System Facilities				
• Limited	2.7	2.7	3.5	2.5
• Unimproved	19.6	9.1	5.6	2.0
• Basic Sanitation	47.0	44.0	20.1	9.6
• Safely Managed	29.5	43.9	70.7	85.6
Total	100.0	100.0	100.0	100.0

Note: Percentages may not total 100% because of rounding.
Source: World Health Organization (WHO)/United Nations Children's Fund (UNICEF) Joint Monitoring Programme. Water, Sanitation and Hygiene (WASH). Rural and Urban Sanitation Service Levels, 2015 and 2020: Household Data by Service Level. https://washdata.org/data/household#!/ (accessed 17 May 2023).

---

[18] WHO/UNICEF Joint Monitoring Programme for Water Supply and Sanitation. 2010. *Progress on Sanitation and Drinking Water—2010 Update*. Geneva: WHO.

[19] WHO/UNICEF Joint Monitoring Programme. Water, Sanitation and Hygiene (WASH). Rural and Urban Sanitation Service Levels, 2015 and 2020: Household Data by Facility Type. https://washdata.org/data/household#!/dashboard/new (accessed 17 May 2023).

A programmatic sanitation strategy has two principal objectives: (i) encouraging behavior change at the community-wide level to put an end to open defecation and to improve facilities and practices, and (ii) stimulating demand for sanitation products while concurrently fostering affordable and reliable supplies. Among the programmatic approaches employed to change behaviors, create demand, and increase the supply of goods and services (improving sanitation supply chains) are community-led total sanitation, behavior change communication,[20] sanitation marketing, and sometimes a combination of the three (Appendix 2). They are compatible and complement each other in achieving sustainable facilities and behavior change.[21] For community-led total sanitation, cooperation and social solidarity among community households are vital elements. Behavior change communication integrates best practices from social and commercial marketing.[22] For sanitation marketing, the application of the "Four Ps" (product, price, place, and promotion) augments demand and supply for improved sanitation, especially among poor people.

Operationalizing a programmatic approach for sustainable service delivery at scale requires engagement by all stakeholders—national, provincial, and local governments; the communities or village committees; households; the private sector; and development partners. Table 2.2 summarizes the roles of these key stakeholders. The national and provincial governments will establish and maintain a strong enabling environment, whereas the local governments will manage the implementation. Capacity building should be strengthened for local governments, the private sector, and resource agencies.

Table 2: Components of a Rural Sanitation Service Delivery Model

Stakeholder	Key Role
National/Provincial Government	Create enabling environments for sustainable and large-scale sanitation programs
Local Government	Improve sanitation, establish regulation, define strategies and plans, enable and regulate the private sector, strengthen advocacy and capacity building, and facilitate monitoring and evaluation
Communities/Village Committees	Promote unity for improved sanitation
Households	Arrange, utilize, and maintain improved sanitation facilities
Private Sector	Produce sanitation products and services
Development Partners	Provide technical guidance in terms of planning, design, implementation, and monitoring of rural sanitation programs

Source: Asian Development Bank.

Effective design and implementation of sustainable and large-scale rural sanitation programs require systematic institutional and policy support, strong stakeholder delivery support, and financing strategies that are both affordable and accessible to poor people. Ensuring the supply of sanitation products and services in rural areas can be a critical part of sanitation marketing, overcoming supply

---

[20]   Behavior change communication is research and development of communication materials for promoting positive economic, health, and/or social outcomes. It builds on an understanding of the factors affecting household demand as well as opportunities and constraints in the sanitation supply chain. The evidence needed to develop an effective and sustainable sanitation marketing program often requires primary research, which entails the collection of information directly from the households.

[21]   Many factors can influence behaviors, such as convenience, status, pride, guilt, shame, and well-being.

[22]   For more information, see J. Devine and C. Kullmann. 2011. *Introductory Guide to Sanitation Marketing.* www.wsp.org/sanmarketingtoolkit.

chain constraints and facilitating access to finance for suppliers, as required. Program costs include upfront costs (conducting market research, preparing promotional messages, producing promotional materials, etc.); costs for on-site household sanitation facilities; promotional activities to increase demand and strengthen supply chains; and monitoring and evaluation costs. Details on how to determine the costs of a rural sanitation program are in Appendix 3.

A rural sanitation program is usually developed in stages: (i) situation analysis of the sanitation and hygiene status at the national (and provincial) level, including policies, lessons learned, and implementation capacity; and (ii) program design of results framework and implementation strategy for the program area based on context and conditions, including implementation arrangements, phasing, and capacity development.

Situation Analysis

The situation analysis should inform program objectives and examine the following program areas. Once the program area is determined, the physical and economic conditions can be assessed.

- ***Sanitation and hygiene.*** Review the (i) rates of open defecation;[23] (ii) access to and use of existing sanitation facilities (including handwashing facilities) and safely managed sanitation services; and (iii) environmental sanitation conditions (food hygiene, liquid and solid waste management, etc.).
- ***Water supply.*** Given that water is needed for handwashing, toilet flushing, anal cleansing, and other hygiene activities, lack of water supply poses a major constraint to sustained rural sanitation and hygiene practices. The water supply's condition also affects the amount of fecal sludge produced and, therefore, influences program design and the choice of program area.
- ***Nutrition and health.*** In rural areas, poor nutrition and public health are often linked to inadequate sanitation.
- ***Poverty.*** Data on income poverty (wealth) and basic service deficiencies are usually available from surveys.
- ***Gender and disadvantaged groups.*** Disaggregated data on the sanitation, hygiene, and health status of women and girls, as well as other disadvantaged and vulnerable groups, should be considered in the selection of the program area and program targets.[24]

Program Design

Evidence-based research and lessons learned from previous rural sanitation and hygiene programs form an important basis of a program's design. Behavior changes of the target area's population with respect to the barriers to and the drivers of sanitation and hygiene should be determined, as should preferences for specific services, willingness to pay, and expectations.

---

[23] For example, open defecation, together with poor hygiene practices, is linked with child growth issues such as stunting.

[24] Development partners, such as UNICEF, promote the use of an integrated (or convergence) approach, where the rights of children are all addressed simultaneously to ensure comprehensive outcomes for them.

Assessing the enabling environment. An enabling environment refers to a set of conditions that supports the effectiveness and sustainability of the institutions, systems, and outcomes of rural sanitation.[25] The Sanitation and Water for All global Partnership, a multi-stakeholder global platform to support member countries in achieving universal access to WASH, promotes the following five building blocks of a well-functioning water, sanitation, and hygiene sector: (i) sector policy and strategy; (ii) institutional arrangements; (iii) sector financing; (iv) planning, monitoring, and review; and (v) capacity development.[26]

Program objectives and targets. National priorities should guide programs and their targets. However, the program objectives and targets need to be realistic—they should reflect what is possible during the program timeframe, given the potential challenges in the program area. The program design should allow for an adaptive management approach.

Capacity development. Good program management is vital in ensuring program effectiveness. It requires giving particular attention to the program appraisal and design stages. Capacity assessment involves the identification of existing capacity (actors, institutions, systems) and highlights any key capacity gaps which need to be filled through capacity development. The institutional model for program management should be decided early in the program design process. It is likewise useful to include in the program results framework (objectives and targets) specific equity and sustainability targets to promote greater attention to these areas during program implementation as well as program monitoring and evaluation.

Monitoring and evaluation. An overall program monitoring and evaluation framework should be determined during the program design stage. This should include the main intended results, indicators, and targets (annual and total). Capacity development for sustainable support and monitoring may be needed, informed by a program capacity assessment.

---

[25] D. Tsetse et al. 2016. Strengthening Enabling Environment for Water, Sanitation and Hygiene. *Guidance Note*. New York: UNICEF. https://www.ircwash.org/resources/strengthening-enabling-environment-water-sanitation-and-hygiene-wash-guidance-note.

[26] The strength of the enabling environment for rural sanitation and hygiene will influence the program design for an area. If the enabling environment is strong, then the implementation strategy should be comprehensive, adequately covering the program area, addressing inequalities, and developing effective approaches for sustainability. If the environment is weaker, then the implementation strategy needs to be more focused, targeting high-priority areas (particularly those where supportive partners are available) and only scaling up after initial implementation is achieved.

INSTITUTIONAL SETTING

The Government of the PRC prioritizes the treatment of rural domestic wastewater as part of improving the rural living environment. In 2018, the government issued a Letter about the Division Scheme for Rural Living Environment Improvement which stipulates the division of responsibilities among the national ministries and agencies involved in promoting rural wastewater treatment management. These include MARA, MEE, MHURD, Ministry of Finance (MOF), Ministry of Health, Ministry of Science and Technology, Ministry of Water Resources, and NDRC; their roles and responsibilities are presented in Table 3.1. The Central Agricultural Office (CAO), MARA is the lead entity. Regarding fund management, the NDRC and other ministries have, in recent years, issued guidance to local governments on areas to improve their cost recovery strategies and policies for wastewater treatment and establish a farmer payment mechanism.

There is currently no consistent framework in the PRC for rural wastewater treatment, indicative of the state of leadership fragmentation. RWWM is guided vertically by ministries, with implementations led by the subnational (provincial and local) governments. The national government plans overall, provinces assume overall responsibility, and cities and counties focus on implementation and operations.

Table 3: Roles and Responsibilities for Rural Wastewater Management in the People's Republic of China

Functional Responsibilities	Responsible Ministry/Agency							
	NDRC	MOF	MHURD	MARA	MEE	MST	MWR	MOH
General Functions								
Strengthen the construction of wastewater treatment facilities and the technological support for the treatment.			X		X	X		
Incorporate rural water environmental governance into the river chief and lake chief management systems of the PRC.							X	
Promote local cleanup of village water bodies and gradually eliminate black and odorous water bodies in rural areas.			X	X	X		X	
Funds Management Functions								
Improve the construction and management mechanism.	X	X	X	X[a]	X			X[b]

continued on next page

Table 3, *continued*

Functional Responsibilities	Responsible Ministry/Agency							
	NDRC	MOF	MHURD	MARA	MEE	MST	MWR	MOH
Define and improve price mechanisms.	X							
Manage the rural environment improvement funds.		X			X			
Formulate fund allocation standards, review plans for allocation of funds, issue budgets, implement a budget performance management process, strengthen supervision of the use of funds, and guide local budget management.		X						
Guide the implementation of rural environmental improvements (and organize and develop related reserves), propose tasks and fund allocation formulas, conduct supervision and evaluation, and promote and guide local governments on good budget performance management.					X			

MARA = Ministry of Agriculture and Rural Affairs, MEE = Ministry of Ecology and Environment, MHURD = Ministry of Housing and Urban–Rural Development, MOF = Ministry of Finance, MOH = Ministry of Health, MST = Ministry of Science and Technology, MWR = Ministry of Water Resources, NDRC = National Development and Reform Commission, PRC = People's Republic of China.
[a] Together with the Central Agricultural Office.
[b] Together with the Health Commission.
Source: Asian Development Bank (ADB). 2022. *Rural Wastewater Management in the People's Republic of China*. Consultant's report. Manila (TA 9825-PRC).

National Level

Rural wastewater treatment is prioritized as a key task in the National Rural Vitalization Strategy. According to the Guiding Opinions on Promoting Rural Wastewater Treatment, issued by nine ministries in 2019, rural wastewater treatment institutions in the PRC include institutions at the central, provincial, city, and county levels.

The development of rural wastewater treatment in the country is carried out under the management of the Central Committee of the CPC and the State Council at the national level; and the provincial, municipal, county, and township party agencies and local governments. The key ministries involved are MHURD, MARA, and MEE. They are responsible for improving rural living environments and guiding rural ecological environmental protection.

A Central Leading Group for Rural Work coordinates rural and agricultural matters (including policies, funding, sector issues, and human resources) and supervises and promotes implementation. For rural wastewater treatment, this leading group coordinates with the MHURD, MARA, MEE, and NDRC as required. The government ministries issue supporting policies. The MEE focuses on planning, standards, technical support, and pilot demonstration projects. The NDRC has the primary responsibility for tariff and cost recovery policies, and the MOF is in charge of the rural environment improvement fund, central government funds, and other financial support. MST is responsible for guiding rural scientific and technological progress, while the NDRC takes care of system reforms.

The Central Leading Group for Rural Work is the coordinating body of the Central Committee of the CPC and is headed by the Vice Premier of the State Council. The office of the leading group (referred to as the CAO) is in the MARA. The head is the secretary of the Party Leadership Group of the MARA. The CAO organizes and conducts policy research on issues relating to agriculture, farmers, and rural areas. It coordinates and supervises relevant parties to implement decisions and directives issued by the leading group.

Local Level

At the local level, provincial party committees and governments are responsible for improving the rural living environment in their respective provinces. This includes treating and managing rural wastewater, establishing policies and procedures, and improving organizational leadership. The city and county party committees and governments focus on project implementation, the effective use of funds, and the O&M of systems and facilities. The township party committees and governments are responsible for the actual implementation of projects within their jurisdictions. The village party organization is responsible for supervising and for enhancing farmers' awareness of environmental protection practices. Throughout government levels, the agriculture and rural departments take the lead in improving the rural living environment, and the ecological environment departments are responsible for rural wastewater treatment.

4 POLICIES AND REGULATIONS

National Laws and Regulations

The Government of the PRC has always given great importance to rural areas. For 18 consecutive years, from 2004 to 2021, the No. 1 Central Document of the country emphasized issues relating to agriculture, rural areas, and farmers.[27] However, because of the PRC's vast territory and significant differences between regions, rural wastewater treatment is still in the process of early development in most provinces. Ministries and commissions focus on high-level systems design and regulate emission standards. For implementation, starting with demonstration cases and proceeding in stages is emphasized. Building on the 13th Five-Year Plan (2016–2020), the 14th Five-Year Plan (2021–2025) is focused on creating a long-term development mechanism.

There is no specific national law on rural wastewater in the PRC. The Water Law (2002) sets out to "rationally develop, sustainably utilize, preserve, and protect water resources to satisfy the needs of the urban and rural inhabitants in their domestic use of water." The law gives overall consideration to the agricultural, ecological, and industrial needs for water; however, it does not have specific requirements for rural domestic wastewater. The Water Pollution Prevention and Control Law (amended in 2017) emphasizes that "the state supports the construction of rural wastewater and garbage treatment facilities and promotes the centralized treatment of rural wastewater and garbage," and that "the local governments at all levels should make overall planning to build rural wastewater and garbage treatment facilities and ensure their normal operation."

Regarding rural environmental protection, the national government has, in recent years, issued a series of policies on rural wastewater treatment (Appendix 4). Due to the differences between the provinces, the central government adopts the principles of (i) promoting approaches according to local conditions, and (ii) proceeding step by step. Additionally, the national government has issued a series of standards and regulations for the treatment of rural domestic wastewater.

---

[27] The No. 1 Central Document is the first policy statement released by the government each year, which serves as an indicator of the government's policy priorities. It is published by Xinhua News Agency.

The Rural Vitalization Strategy, 2018–2022

Improvement in rural livelihoods, including agricultural production, has lagged behind that of urban livelihoods and industrial production in the PRC. This has resulted in the persistence of poverty and income inequality in rural areas, particularly in the less developed regions. Most rural wastewater is directly discharged into the environment without appropriate treatment and disposal. Against this backdrop, the Government of the PRC launched in 2017 the ambitious National Strategic Plan for Rural Vitalization for 2018–2022 (the rural vitalization strategy) at the 19th National Congress of the CPC. The strategy aimed to promote innovative approaches for rural development, poverty reduction, improved ecological practices, and green and inclusive growth.

Among the strategy's focal areas were rural waste and wastewater management, agriculture sector modernization, ecosystem services protection, rural–urban integration, rural education and health services, and rural governance reform. The strategy aimed to modernize the rural economy and address urban–rural disparities through improvements in infrastructure, promotion of supply-side structural reform in agriculture, innovations in science and technology, strengthening of rural land ownership, and reforms in the subsidy system. Sound institutional systems, policies, and mechanisms for integrated urban–rural development were also established. [28]

In April 2021, the Rural Vitalization Law was adopted by the PRC's national legislation, enshrining in law a legal responsibility to pursue fervently the rural vitalization strategy.[29] Moreover, the government's No. 1 Central Document for 2021 charts a road map for the rural vitalization strategy. It included targets and tasks for the year in relation to the agriculture sector and rural areas, highlighting the important role of science and technology. Besides boosting rural local industries, the government attached great importance to improving the rural living environment and ensuring better access of residents in rural areas to improved services and facilities. The government will gradually shift the policy focus from poverty alleviation to comprehensive promotion of rural vitalization.

Important progress has been achieved on rural vitalization. In 2020, a robust institutional and policy framework was created. According to the MARA, industries in rural areas experienced steady growth in the first half of 2021. The "toilet revolution" was placed high on the national agenda to ensure a more livable countryside.[30] From 2018 to 2020, official statistics reported that over 40 million rural household toilets had been renovated, achieving a 68% penetration rate of clean toilets in the PRC's rural areas.[31] The rural toilet revolution task—in many cases, conversion of dry toilets to bathrooms with flush toilets and shower fittings—continues, as do efforts to improve rural household solid waste and sewage treatment.

---

[28] Xinhua News Agency, based on a speech by President Xi Jinping at the Eighth Group Study Session of the Political Bureau of the 19th CPC Central Committee. 21 September 2018.

[29] The law came into force on 1 June 2021.

[30] The toilet revolution refers to a government campaign launched in 2015 aimed at improving the sanitation and conditions of public toilets in the PRC's tourist areas and ensuring villagers' access to hygienic toilets. Between 2015 and 2017, the government constructed over 68,000 public toilets in the PRC.

[31] Y. Gao et al. 2022. Assessment of Environmental and Social Effects of Rural Toilet Retrofitting on a Regional Scale in China. *Frontiers in Environmental Science*. 10. https://doi.org/10.3389/fenvs.2022.812727.

Local Regulations

To further improve and standardize rural domestic wastewater treatment and strengthen O&M management, most provinces have introduced management and policy requirements. Appendix 5 lists recent provincial policy documents on the subject.

The PRC's first provincial-level local regulation in the field was issued in January 2020 in Zhejiang Province.[32] This regulation stipulates the scope of treatment and treatment facilities, the construction and supervision responsibilities of government departments, the conduct of O&M units, particularly related to the facility users, and the legal responsibilities of involved parties. The Zhejiang regulation is a useful reference for other provinces.

As mentioned, for rural wastewater treatment, the central government follows the principles of promoting according to local conditions and proceeding step by step. During the 13th Five-Year Plan (2016–2020) period, the focus was on starting with demonstration cases and promoting the subject gradually. Ministries and commissions focused on high-level systems design and regulated discharge standards. The 14th Five-Year Plan (2021–2025) focuses on problems such as unbalanced regional development, imperfect basic living facilities, and weak management and maintenance mechanisms. It emphasizes the prioritization of local planning and the importance of a long-term development mechanism.

Rural wastewater treatment in the PRC still faces practical problems such as insufficient local attention and the difficulty of changing people's habits. Although some provinces have issued local regulations for rural domestic wastewater, there is no specific legal support on the subject. Only by incorporating the treatment of rural domestic wastewater into the rule of law can a beautiful countryside and the full implementation of the rural vitalization strategy be ensured.

---

[32] Provincial Government of Zhejiang, PRC. 2020. *Regulations on the Management of Rural Domestic Wastewater Treatment Facilities in Zhejiang Province.* Hangzhou.

5
FINANCING AND MANAGEMENT OF RURAL WASTEWATER SERVICES

The construction, operation, maintenance, and supervision of facilities require long-term and significant continuous investment. Construction and operation still require financial input from governments at all levels in the PRC. Investments by the private sector in rural wastewater treatment largely need to rely on long-term support from public finance to achieve basic investment returns.[33]

A locally based, centrally subsidized, and participatory fundraising mechanism exists for rural wastewater treatment in the PRC. The national government is increasing investments in rural domestic sewage treatment, prioritizing poverty areas to some extent. Rural areas that have centralized wastewater treatment facilities are encouraged to explore a cost-sharing mechanism with farmers. In some provinces, funds for the management of treatment facilities are included in the city and county budgets. The central government, mainly through the MOF, arranges financial support for rural wastewater treatment (centralized subsidies). In June 2021, for example, the MOF issued the Measures for the Management of Rural Environmental Remediation Funds, which stipulates the division of responsibilities for the rural environment improvement funds.[34]

Sources of Funds

The primary sources of funds for rural wastewater treatment and improvement of the rural living environment in the PRC are summarized as follows based on the Five-Year Action Plan for the Improvement of Rural Living Environment (2021–2025). At present, the primary source of funds

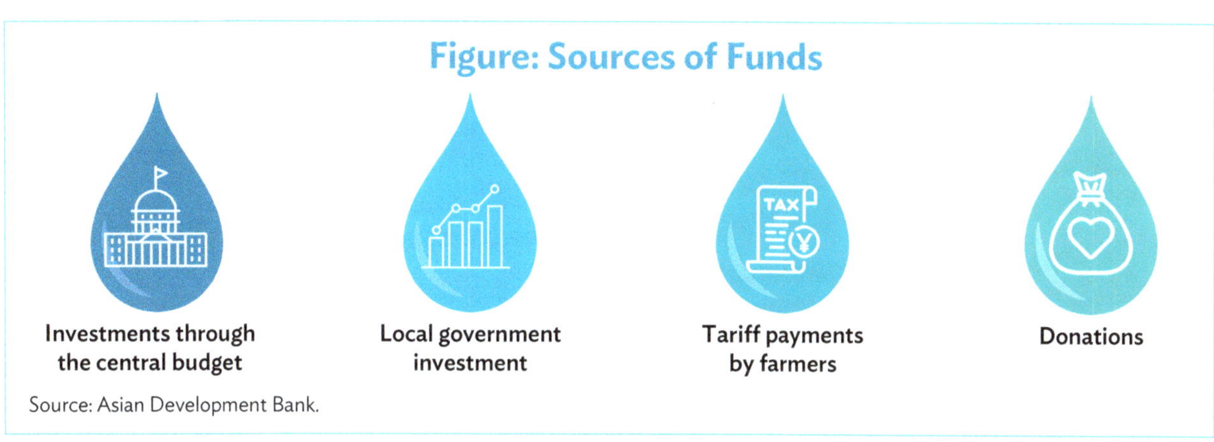

Source: Asian Development Bank.

---

[33] "Private sector" refers to the term "social capital" commonly used in Chinese documents.

[34] The rural environment improvement funds are arranged by the general budget of the central government to support local implementation of rural ecological environmental protection work and promote the improvement of rural ecological environment quality.

is local government investment, with some rewards and subsidies from the central government. User tariffs for farmers are being piloted in some areas where conditions permit, but this will likely not become the primary source of funds for rural wastewater treatment in the long-term.

Investments through the central budget. Reward and subsidy contributions from the central government complement local government funds. To ensure the effective use of national rural environment improvement funds, the action plan aimed to "improve the local government investment mechanism with appropriate incentives and subsidies from the central government, continue to arrange investment through the central budget, implement the rural toilet revolution as planned, and promote financial incentives and subsidies for the village." To respond to the relatively weak rural infrastructure in the central and western regions of the country, the NDRC and MARA jointly launched in 2019 the Rural Living Environment Improvement and County Promotion Project in the national investment budget. During the first two years, this project invested CNY6 billion (or close to $1 billion by mid-2022) to support the construction of rural solid waste and wastewater treatment facilities and other infrastructure in poor areas. In addition, the MEE has implemented a policy of 'promoting governance with rewards' in rural areas. During the 13th Five-Year Plan (2016–2020), CNY26 billion ($3.8 billion) was allocated by the PRC to make environmental improvements in numerous villages, including the support of rural wastewater and garbage treatment and drinking water source protection.

Local government investment. The action plan states that local governments at all levels should ensure investment and O&M funds for improved rural living environmental infrastructure. It encourages local governments to collect land transfer fees, issue bonds, and use other means to improve the rural living environment. Although macro-control measures may constrain these funding channels, these are still important for local rural environmental governance. County-level governments can integrate funds and projects to improve the rural living environment on a village-by-village basis. This includes agriculture-related funds arranged at the county level. Some local governments have included the cost of rural wastewater treatment operations in their budgets and increased their investments.

Tariff payments by farmers. Implementation of a "user pays" system is a solid solution to improve the rural living environment. However, considering the capacity of farmers to pay, this needs to be established gradually. For rural wastewater treatment, it may for some time be limited to O&M costs.

Donations. Individuals, enterprises, and social organizations may be attracted to support the improvement of the rural living environment by donating funds or materials, or through in-kind assistance. However, this continues to be only a limited source of funds.

Social Capital

Governments at all levels in the PRC encourage the participation of social (private) capital in the rural wastewater treatment sector. At present, the biggest problem faced in attracting this participation is that it is often difficult to ensure an adequate, long-term, and stable investment return. Some local governments have issued policies to encourage this through a public–private partnership model, which combines wastewater treatment investments with other local infrastructure investments (e.g., solid waste, power, etc.), making the total package more attractive to a private investor.

Governments also encourage commercial financial institutions to provide credit support for rural wastewater treatment investments, but they face similar challenges as social (private) capital. Due to a lack of reasonable capital return expectations, this is still at an exploratory or piloting stage.

At the village and county levels, education programs are critical to ensure adequate community knowledge about sanitary practices and institutional and technical capacities to manage wastewater facilities.[35] Effective participatory processes and public awareness are important for local communities so they may have a voice in the development and operation of wastewater management in their village. Participation is defined here as a process whereby key stakeholders influence and share management of their development initiatives, decisions, and resources. Under this definition, stakeholders avoid being treated merely as passive and assisted recipients, interview objects, or labor during development. It is a process that stimulates stakeholders to participate at every project stage.

Effective participation needs the cooperation and support of the primary stakeholders (i.e., direct beneficiaries and any adversely affected people or organizations) as well as the other stakeholders (including the project owner, construction and consulting units, project implementation entities, and relevant government agencies or departments).

Community participation. Consultation with project stakeholders helps promote ownership and active engagement of the communities (photo by ADB).

---

[35] This section refers to wastewater management systems in rural areas for which the ownership and management is transferred to the villagers after project completion. The section draws on the Ningbo Rural Development Project Social Assessment Report—Community Participation Manual, prepared by the Ningbo City Rural Domestic Wastewater Treatment Project Office in June 2009 under an investment project financed by the World Bank.

A participation process should be based on the following principles:

- ***Be inclusive***—to have hearings and discussions to understand the needs and ideas of the participants for the purpose of reaching an agreement.
- ***Be transparent***—information should be provided and obtained in an open way and be publicized. The content and feedback should be clear and concise.
- ***Be cooperative***—the stakeholders should work together to seek mutual agreements and results.
- ***Be comprehensive***—the process should create awareness and understanding of views, including opposing views.
- ***Have integrity***—the process must be carried out with mutual respect and integrity.

Rural domestic wastewater treatment, is usually carried out in three stages, during which stakeholder participation is essential: (i) project preparation, (ii) project implementation, and (iii) O&M or ongoing management (Table 4). The preparation stage includes project planning until project launch and the start of construction. Implementation is usually defined as the period from project launch until 18 months after the final work acceptance, including equipment commissioning and trial operation. The O&M stage is from the time property rights of facilities and networks are officially transferred to the village committee(s).

Table 4: Stakeholder Participation in a Rural Domestic Wastewater Program

Stage	Key Tasks	Consultants, Contractors, and Operations Unit	The Community and Other Stakeholders	Government Units
Preparation Stage	• Decide on village(s) and site(s) • Make preliminary designs • Prepare construction site(s)	• Community orientation/training • Consultant proposes village(s) and designs for the initiative, and discusses with the village committees and other stakeholders	• Provide comments on the village selection, the site(s), and the design(s)	• Decide on village(s) and carry out publicity • Complete land acquisition and resettlements, as needed
Implementation Stage	• Detailed design of facilities and pipe networks • Construction	• Contractors provide relevant construction information to the community • Training of community personnel for the system operations	• Provide comments	• Determine personnel for the system operations • Training of personnel
O&M Stage (Ongoing Management)	• System operation and management	• Operating company performs in accordance with the requirements for the wastewater treatment system (as per contract)	• Community participation in the O&M of the system	• Transfer of ownership • Training of the community • Engage a system operator • Supervision and inspection of the operations

O&M = operation and maintenance.
Source: ADB. 2022. *Rural Wastewater Management in the People's Republic of China*. Consultant's report. Manila (TA 9825-PRC).

Grievance and appeal of villagers. Villagers can convey any grievances to the village committee. If the villagers do not accept the decision of the village committee, they can appeal to the local government. If they do not accept their decision, they can either appeal to the next government level, or the civil division of a people's court, according to civil law in the PRC.

7

INTERNATIONAL EXPERIENCE AND GOOD PRACTICES

This section highlights some good practices related to rural sanitation development based on international experience.

Prerequisites for Effective Rural Sanitation

The Sustainable Development Goals of the United Nations call for universal access to water and sanitation services, ecosystem management, and enhanced sanitation to improve health. At present, approximately 4.5 billion people globally do not have access to safely managed sanitation services, while 2 billion people are without access to safely managed drinking water services. In addition, unsafe hygiene practices compound the resulting impacts on people's health.

The following five aspects are key for effective rural water and sanitation services.[36]

- **Institutions.** Institutional arrangements must provide effective incentives and adequate resources, and the organizations that deliver services need to have the requisite capacity. The quality and sustainability of sanitation services are bound by the rules within which these organizations operate. To strengthen institutions and service accountability, change is required and should be grounded in the local context (e.g., in the local culture, economy, political circumstances, etc.).
- **Financing.** For the PRC to meet the Sustainable Development Goals by 2030, financing needs are estimated to be in the trillions of dollars. A two-pronged approach is needed: (i) improving the financial viability of the water supply and sanitation sector through cost recovery mechanisms, while ensuring affordable services for poor people; and (ii) leveraging commercial and other nonstate sources of financing.
- **Sustainability.** This is about ensuring that available resources can continue to deliver services to future generations. This includes: (i) sustainable management of water resources to secure their long-term availability; and (ii) adequately built water supply and sanitation infrastructure assets.
- **Inclusion.** Universal access to water and sanitation services includes improving the situation for disadvantaged individuals and groups. This requires knowledge about the nature of service inequalities, enhancement of capacities, and establishment of incentives for better outcomes. Such measures demand institutions to hold service providers accountable.
- **Resilience.** Resilient solutions need strategies that incorporate climate and disaster risk considerations (to save lives and livelihoods) and respond to socioeconomic and environmental considerations.

---

[36]   This draws on material from the Global Water Security and Sanitation Partnership at the World Bank.

International Good Practices on Rural Development

During the Rural Vitalization and Modernization session at the 2018 China Development Forum in Beijing, organized by the China Development Research Center of the State Council, the International Food Policy Research Institute shared success stories, lessons, and strategies from the European Union, Japan, the Republic of Korea, Thailand, and the United States.

For the implementation of the PRC's rural vitalization strategy, the following key lessons can be taken from the experiences of these countries:

- Environmental issues in wealthier rural areas can be addressed by focusing on collective action, behavior change, and education on environmental health linkages.
- In poorer rural areas, emphasis should be on upgrading infrastructure, improving access to quality education, and facilitating migration.
- Rural communities and villages should be empowered and motivated by promoting local production of niche products for healthy foods from the rural areas.
- Subsidy policies need to be reformed through conversion of production subsidies to income support, livelihoods, and/or improved environment.
- Improving value chains, physical connections, and broadband connectivity can help strengthen rural–urban linkages.
- Agriculture and rural development strategy, policy, and investment should be integrated.

The **European Union** has been converting agricultural subsidies to rural development support and linking payments to sustainable agriculture, food safety, and environmental protection. Production subsidies have shifted to direct payments and income support. Meanwhile, special funds to support rural areas have been set up. The new European Union farm policy has also introduced more equitable and better-targeted safety nets, which reward farmers for sustainable and climate-smart practices.

To connect local farmers to urban consumers, **Japan** has promoted inclusive rural–urban linkages through investments in rural infrastructure, farmers' markets, and cooperatives. Like the PRC, Japan's rural populations are becoming older as young people migrate to urban areas. Hence, in 2000, Japan implemented a mandatory long-term social care insurance program, which aims to provide affordable home- and community-based services (e.g., home assistance, adult day care, and visiting nurses) to senior citizens in rural areas.

The "New Village Movement," launched in the 1970s, brought substantial changes to the **Republic of Korea** by stressing the importance of self-governance and cooperation of rural communities in modernizing the economy, strengthening infrastructure, improving living conditions, and protecting farmers' rights.

In **Thailand**, the development of niche products and the empowerment of rural areas are key elements of poverty reduction initiatives. The government's "One Tambon One Product" program, which supports the production and marketing of local goods from rural areas, not only encouraged local entrepreneurship but also provided alternative income to farmers. To increase local employment and sustainable livelihoods, Thailand promoted other rural-based initiatives, such as rural tourism, organic rice farming, and handicraft production.

In 2018, the **United States** unveiled a $1.5 trillion infrastructure investment program to improve rural infrastructure, including transportation, safe drinking water, renewable energy, and broadband connectivity. New technologies (e.g., e-commerce) and work modalities (e.g., telework) have also shown potential to revitalize rural areas by creating income-earning opportunities and keeping rural towns connected.

Private Sector Participation

Experiences from developing countries have demonstrated how the ability of communities to manage rural piped water and wastewater schemes can be limited by technical complications, a relatively small customer base, or large financial requirements for investments and O&M. Additional challenges include subsidies required for capital costs, issues with tariff collection, volatile local government commitments, and lack of capacity (professional or entrepreneurial) to effectively operate and manage schemes once constructed.

A private sector participation approach often appears to be the best option to ensure the sustainability of a system. However, sustained private sector participation can also be a challenge because of unpredictable community demand, the reluctance of users to pay tariffs, low profitability for the investor, and the inherent disincentive for long-term investments due to these risks. Approaches to mitigate these risks include: (i) clustering multiple schemes to attract bigger and more capable private sector participants; (ii) shorter operational contract periods, with options of contract renewal; (iii) stronger partnership and clarity of roles between the private investor, community, and local government; (iv) active facilitation of the operator–consumer relationship; (v) constant monitoring; and (vi) financial and technical assistance for business.

A case study of Changshu County in Jiangsu Province provides the first comprehensive demonstration area for integrated rural wastewater governance in the PRC based on public–private partnership projects for decentralized county wastewater treatment. This case study proposes (i) to thoroughly take into account the characteristics of decentralized wastewater treatment in rural areas, the current industrial technology level, and the rural community's investment capacity; and (ii) to systematically formulate water quality standards for rural wastewater treatment according to the 3A principle—appropriateness, availability, and affordability.[37] It highlights the important role of public–private partnerships in ensuring the efficiency and quality of rural wastewater treatment projects.

---

[37] B. Fan et al. 2022. *Public–Private Partnerships for Wastewater Treatment in Rural Areas: Case Study of Changshu, People's Republic of China.* ADBI Development Case Study No. 2022-1. Tokyo: Asian Development Bank Institute.

8

THE WAY FORWARD: CONCLUSION AND RECOMMENDATIONS

Conclusion

The Government of the PRC attaches great importance to rural wastewater management, as reflected in its recent 5-year plans.[38] The national policy on rural wastewater treatment adheres to the principle of adapting measures to suit local conditions and progressing step by step. It established a comprehensive management structure in coordination with the CAO in the MARA and issued a series of policies and regulations to guide the country on the subject. The promotion of rural wastewater treatment has achieved considerable results during the last 5 years. However, from the perspective of legislation, the PRC has no specific law covering rural domestic wastewater. Instead, some leading provinces have issued local regulations as legislative support for rural domestic wastewater treatment in their respective regions.

The PRC has an established decision-making and coordination mechanism for promoting rural wastewater treatment. Strategies are set at the central level, while the implementation is accomplished at the local level. Local governments may, however, follow a local institutional arrangement different from that at the central level. This mismatch between institutional arrangements at the local and central levels may lead to confusion and difficulty in policy coordination over the long run.

Many policies and regulations issued by the central and local governments in the country have followed a step-by-step pattern, with demonstration cases in the better-prepared regions occurring before upscaling and replication. Although all provinces have established local discharge standards, many of them tend to be excessively strict. This may not permit or encourage the practitioners to adopt the most cost-effective solutions for wastewater treatment.

There are substantial differences in the management level of rural domestic wastewater treatment across the country. For example, Zhejiang Province has formed an advanced policy system with many regulations issued in relation to the construction, technical transformation, management, and O&M of its rural domestic wastewater treatment. Some provinces are still in the planning stages of facility construction. However, even in the most advanced provinces, it is difficult to enforce the "polluter pays principle" and adequate cost recovery from the users at this development stage. For private sector engagement, the main challenge remains achieving a stable mechanism for the sector to get an appropriate return on their investments.

---

[38]   The 13th Five-Year Plan (2016–2020) focused on demonstration first, construction of facilities in stages, and strengthening a high-level design program. The 14th Five-Year Plan (2021–2025) strives to build a systematic, standardized, and long-term policy system and development mechanism.

Recommendations

A comprehensive approach to wastewater management in rural villages includes institutional, regulatory, technical, and financial matters, as well as educational and participatory activities. It requires engagement at the village, county, municipal, provincial, and central government levels. The institutional and regulatory aspects include the establishment of clear and effective policies, regulations, and governance arrangements to support the planning, implementation, O&M of facilities, and the management of related development projects. Appropriate sanitation technologies at a reasonable cost need to be selected, and they need to be easily and reliably operated and maintained, ideally by the villagers themselves. The financing of projects will likely use public funds and subsidies for quite some time to assist low-income villagers, with some contributions by the beneficiaries (financially or in-kind). The costs of O&M need to be covered, at least partly, through a cost-sharing arrangement between villagers and the local government.

The following initiatives and actions are recommended for the PRC:

Institutional

- To ensure the effective implementation of national and local rural wastewater treatment policies, the division of labor needs to be further streamlined among ministries at the national and local levels, led by environmental departments. The local organization should ideally be consistent with the one at the central government level.

Policies and regulations

- Special laws and regulations should be formulated to bring the subject of rural wastewater treatment into the legal system and ensure its advancement in conjunction with the accelerated implementation of the country's rural vitalization strategy.
- The rural wastewater discharge standards of all provinces and localities need to be reviewed to ensure that they are appropriate and adapted to local circumstances, and not unnecessarily strict.
- The overall design framework, from facility construction through to supervision of O&M, should be refined.
- Assessment of the effectiveness of rural wastewater treatment processes and procedures needs to inform the continuous development and refinement of local policies.

Financing

- More pilot and/or demonstration projects should be implemented to explore new and innovative technologies, cost-effective solutions, and financing models for the construction and O&M of rural wastewater treatment facilities, considering local conditions and potential environmental improvements. This should include models for attracting and managing private sector participants and for appropriate cost recovery from the villagers through user charges.

Stakeholder participation

- Project stakeholders should be involved in the decision-making process through a consultative approach from the start to the completion of an RWWM project. When the voices of stakeholders are heard, it ensures that project plans and designs reflect the actual needs and priorities of the community. Stakeholder participation not only promotes transparency and accountability but also fosters ownership by the stakeholders, which will ultimately lead to improved project sustainability.

SUMMARY OF PROVINCIAL CASES

The provinces in the PRC are different in terms of geography, economic development, and rural domestic wastewater development. The situation in Jiangsu Province, Shanxi Province, and Zhejiang Province are described in terms of their institutional arrangements, policies, and regulations on the subject.

Jiangsu Province

In 2007, Jiangsu launched a pilot project of rural wastewater treatment in the Taihu Lake area, and then gradually rolled it out across the province. By the end of 2020, 11,000 administrative villages in the province had built treatment facilities. The coverage rate of natural villages was 31%, and the coverage rate of rural households was about 30%.

The Action Plan for the Improvement of Rural Wastewater Treatment in Jiangsu Province in the provincial 14th Five-Year Plan (2021–2025) made comprehensive arrangements for rural wastewater treatment. By 2025, the coverage rate of rural wastewater treatment in natural villages in southern Jiangsu and other prioritized areas is estimated to reach 90%, and the coverage rate in administrative villages in central Jiangsu and northern Jiangsu is set to reach 80%.

Sihong County in Suqian City (a typical case). This county is in the west of the province, with an area of 2,731 square kilometers. It has jurisdiction over 12 towns and 4 townships. By the end of 2020, the county had built 150 township and village treatment plants with 320 kilometers of networks, and 400,000 square meters of ecological wetlands. The coverage rate in townships and towns reached 100%, and in administrative villages, it reached over 75%.

Both household treatment and relatively centralized treatment are being used. The main household treatment method is an anaerobic/aerobic(A/O) contact oxidation process, and the relatively centralized facilities use A/O and A/A/O contact oxidation processes. If there is a requirement for total phosphorus removal, artificial wetland or chemical phosphorus removal technology is used, depending on the local situation.

The Sihong Ecological Environment Bureau is the lead department for rural wastewater treatment in the county. They mainly entrust third-party (private) enterprises for their operations. Sihong County Housing and Urban–Rural Development Bureau is the administrative department for urban water supply and sewage treatment in the county. They entrust the local water company to collect the rural wastewater treatment fee.

Sihong County raises funds from the government and the public and private sectors to ensure the normal operation of their rural wastewater treatment facilities. The O&M costs include personnel costs, laboratory fees, facility and equipment maintenance fees, water quality monitoring fees, sludge treatment and disposal fees, and electricity fees. In 2021, the Sihong County Development and Reform Bureau, the Housing and Urban–Rural Development Bureau, the Water Conservancy Bureau, and the Finance Bureau formulated a notice levying a wastewater treatment fee for rural residents and nonresidential users. A subsidy allocation for the O&M of the facilities is included in the annual county budget. Enterprises, social groups, and individuals are encouraged to participate in the construction and renovation of the treatment facilities through investments and donations. To make adequate profits, the operating enterprises may participate in the development and operation of other utility network (e.g., water, gas, and power).

Shanxi Province

Policies issued by Shanxi Province on rural domestic wastewater treatment are listed in Table A1.1. Among them, the "Measures for the Operation and Management of Rural Domestic Wastewater Treatment Facilities in Shanxi Province (for Trial Implementation)," was issued in October 2020:

- to standardize and strengthen the O&M of Shanxi's rural domestic wastewater treatment facilities, increase their operational efficiency, and improve the environment;[1]
- to clarify the responsibilities of the provincial, municipal, and county eco-environmental departments and other relevant departments, county-level governments, township governments (for management), village-level organizations (for implementation), any third-party O&M agencies (as service providers), and farmers (as main beneficiaries);
- to entrust a professional company, the township (town), or the village to conduct the O&M and management of facilities;
- to establish a supervision and assessment system to ensure normal operation of facilities; and
- to encourage the county-level government to include the funds required to operate the treatment facilities in their annual budget and establish a diversified capital investment mechanism to raise funds from multiple sources.

Table A1.1: Policy Documents on Rural Domestic Wastewater Treatment in Shanxi Province

Issued	Policy Document Name
2018	Three-Year Action Implementation Plan for the Improvement of Rural Human Settlement Environment in Shanxi Province
2019	Notice on Carrying out Rural Domestic Wastewater Treatment Work (Jinhuan Soil [2019] No. 20)
November 2019	Pollutant Discharge Standards for Rural Domestic Wastewater Treatment Facilities in Shanxi Province (DB14/726-2019)
June 2020	Technical Guidelines for Rural Domestic Wastewater Treatment in Shanxi Province (DB14/T727–2020)
October 2020	Measures for the Operation and Management of Rural Domestic Wastewater Treatment Facilities in Shanxi Province (for Trial Implementation)
December 2020	Plans for rural domestic wastewater treatment for districts and counties in Datong City

Source: Asian Development Bank.

---

[1] The O&M includes daily maintenance and inspection of facilities, equipment testing and maintenance, sludge disposal, effluent quality monitoring, and emergency response.

Typical cases of rural wastewater treatment in cities and counties. Based on the Notice of the Shanxi Provincial Department of Ecology and Environment on the Progress of the Preparation of the Special Plan for Rural Domestic Wastewater Treatment, issued in December 2020, the districts and counties in Datong City, located in the northern part of the province, prepared plans for rural domestic wastewater treatment.

The Datong City case. Datong City includes four districts, six counties, and 1,925 administrative villages. There are 228 villages that treat their domestic wastewater, of which the wastewater of 140 villages is piped to treatment plants in its urban areas or industrial parks. A total of 88 villages have built treatment facilities. The constructions are financed mainly through subsidies from higher-level governments. However, most village toilets (mostly dry toilets) and kitchen wastewater are still discharged to the ground. The coverage rate of rural sanitary toilets was about 30% in early 2020.

In December 2020, the Special Plan for Rural Domestic Wastewater Treatment in Datong City (2020–2030) was issued. The short-term goal (2020–2024) is to control the domestic wastewater pollution of priority suburban villages and towns in sensitive areas, such as drinking water source protection areas. Since 2020, county-wide planning and promotion have been carried out, including household toilet renovation and other environmental improvements. All rural localities have established long-term mechanisms for their domestic wastewater treatment.

The Yanggao County case. Yanggao County is in the northeastern part of Shanxi Province and is part of Datong City. It covers an area of 1,726 square kilometers. Its water resources consist of spring water, river water, and groundwater. With the rapid development of the local economy, the number of residents in the county has increased, as has the wastewater volume. The county's departments cooperate closely at all levels to promote and improve the treatment of their domestic wastewater.[2] The technology is dominated by biochemical treatment, of which constructed wetlands represent more than 70%. The county prepared the Special Plan for Rural Domestic Wastewater Treatment in Yanggao County, Datong City (2020–2025), which focuses on drinking water sources and water ecological environmental protection. Promotion of the county's rural domestic wastewater treatment is being done step by step. The plan emphasizes the continued strengthening of operations and supervision of existing treatment facilities.

Zhejiang Province

Zhejiang was the first province in the PRC to implement rural domestic wastewater treatment facilities. Since 2003, more than 90% of the 24,000 villages in the province have built such facilities. Since there is no specific legal basis at the national level for the management of rural domestic wastewater treatment facilities in the PRC, the "Regulations on the Management of Rural Domestic Wastewater Treatment Facilities" in Zhejiang Province was issued by the provincial government in September 2020. It makes provisions for the institutional management of rural wastewater treatment in the province. This regulation specifies that:

- The provincial housing and urban–rural construction department is responsible for the guidance and supervision of the planning, construction, renovation, and O&M, as well as the formulation of construction and general discharge standards. The provincial ecological environment department is responsible for the formulation of the water

---

[2]   In 2018, about 93% of the domestic wastewater in the administrative villages of the county is discharged without treatment. Only 15 of the 250 villages (or 6%) have domestic wastewater collection and treatment systems.

pollutant discharge standards for wastewater treatment facilities. The cities and counties are responsible for the supervision and management within their respective administrative areas.

- The departments of agriculture and rural areas, development and reform, natural resources, finance, science and technology, water conservancy, etc., are responsible for the management of treatment facilities in accordance with their respective mandates.
- The township governments and subdistrict or local offices are responsible for the implementation and daily management of facilities in their administrative area.
- The village (resident) committees shall cooperate in the implementation and O&M of the facilities and report to the township or county government on actions that may affect the normal operations of the facilities, including any safety concerns.

The institutional arrangements at the municipal level in the province mirror the provincial-level arrangements. Usually, the municipal housing and urban–rural development department arranges annual assessments of the rural domestic wastewater treatment in the municipality. The local work is assessed on an ongoing basis and scored either as *excellent, qualified,* or *unqualified.* The Municipal Finance Bureau allocates awards based on the assessment results.

Policies and regulations. Zhejiang Province has issued more than 10 relevant laws and regulations covering the construction, technologies, management, and O&M of wastewater treatment facilities (Table A1.2). They provide a reference for other provinces.

Table A1.2: Policy Documents on Rural Domestic Wastewater Treatment in Zhejiang Province

Issued	Policy Document Name
September 2019	Regulations on the Management of Rural Domestic Wastewater Treatment Facilities in Zhejiang Province
January 2020	Technical Guidelines for Online Monitoring System of Rural Domestic Wastewater Treatment Facilities in Zhejiang Province
March 2020	Guidelines for the Diagnosis and Treatment of Common Problems in the Operation and Maintenance of Rural Domestic Wastewater Treatment facilities (Draft for Comments)
March 2020	Wastewater Discharge Standards of Rural Domestic Wastewater Treatment Facilities (DB33/T1196–2020)
April 2020	Guiding Price Guide for Operation and Maintenance Costs of Rural Domestic Wastewater Treatment Facilities in Zhejiang Province (for Trial Implementation)
April 2020	Operation and Maintenance Service Contract for Rural Domestic Wastewater Treatment Facilities in Zhejiang Province (Demonstration Text)
April 2020	Guidelines for the Construction and Maintenance of Sterilization Equipment for Rural Domestic Wastewater Treatment Facilities (Draft for Solicitation of Comments)
April 2020	Technical Regulations for Construction and Renovation of Rural Domestic Wastewater Treatment Facilities
April 2020	Guidelines for the Data Format and Transmission Requirements of the Supervision Service Platform for Rural Domestic Wastewater Treatment Facilities in Zhejiang Province
August 2020	Guidelines for Safe Production Management of Operation and Maintenance of Rural Domestic Wastewater Treatment Facilities
September 2020	Guidelines for the Construction of the Management Platform of the Third-Party Operation and Maintenance Service Organization of Rural Domestic Wastewater Treatment Facilities (Draft for Comments)

Source: Asian Development Bank.

The Regulations on the Management of Rural Domestic Wastewater Treatment Facilities in Zhejiang Province is the first provincial-level local regulation specifically for the construction, management, and O&M of rural domestic wastewater treatment facilities in the PRC. It foresees the following:

- The regulation expands the scope of rural domestic wastewater to include (i) wastewater generated in the daily life of rural villagers, (ii) wastewater generated by public services, and (iii) wastewater generated by activities such as farmhouses.
- It includes facilities that are receiving, transporting, and disposing of wastewater.[3]
- Some rural areas are connected to urban wastewater treatment facilities. In such cases, the local government needs to clarify the main responsible body for O&M, distinguishing private indoor facilities and publicly managed facilities (possibly operated by a private operator or the village collective).
- The code of conduct of the O&M entity should be specified in a service contract, complementing the regulations. Provisions shall, for example, be made for wastewater quantity and quality testing, as well as for any facility damage, malfunction, and outage.
- The regulations follow the principle of "who discharges pays", which are to be established gradually. The tariffs and related procedures shall be formulated by the local government in accordance with relevant national and provincial regulations and the local conditions. The wastewater treatment charges may be combined with the rural domestic water fee.

The Fenghua District case. In 2019, MHURD issued the regulation, County Area Coordination and Promotion of Rural Domestic Wastewater Treatment Cases, describing the Fenghua District in Ningbo City as one of the models for national promotion and reference. In 2016, rural domestic wastewater treatment projects were completed in 137 villages in Fenghua District. Various processes were adopted, such as A/O, anaerobic + land leakage, anaerobic + constructed wetland, trickling filter + constructed wetland, biological turntable, etc. In 2017, the district completed projects in an additional 256 villages. The daily wastewater treatment volume was at that time about 22,000 tons, and the number of households connected was about 76,000. It reached the provincial goal of covering more than 90% of administrative villages. The district established an O&M management system, a remote information system, and an O&M regulatory system. The facilities have a unified design with modular technology. The district entrusted a third party to operate and maintain the facilities.

The Kaihua County case. Since 2013, Kaihua County, a typical county in the province, has actively promoted a new model of rural wastewater treatment, providing a full range of O&M services. Since 2016, it has been rated annually as a pollution control demonstration county. The county lies in the western part of the province and has an area of 2,237 square kilometers and a population of 360,000 (90,400 rural households). It includes eight towns, six townships, and 255 administrative villages. Excluding six villages (to be resettled), in 2017, the wastewater treatment facilities in the administrative villages achieved full coverage. The wastewater network is 1,822 kilometers, covers 255 administrative villages, and benefits 79,000 households. The treated wastewater volume has reached 11,000 tons per day, with a qualified rate of effluent water quality of 100%. All the wastewater treatment facilities are under third-party professional O&M. In 2018, the county made further improvements and expanded the system.

---

[3]   The regulations explicitly exclude treatment facilities built by a single family for self-use or shared by a few families for on-site disposal of domestic wastewater, such as independent septic tanks. These are private operations.

AN OVERVIEW OF RURAL SANITATION APPROACHES

Most rural sanitation and hygiene programs include one or more of the following components:

- community-based behavior change: community-led total sanitation (CLTS);[1] participatory hygiene[2] and sanitation transformation (PHAST),[3] and community health clubs (CHCs);
- market-based sanitation or supply chain strengthening;
- hygiene behavior change communications; and
- systems strengthening, including water, sanitation, and hygiene governance.

In this section, approaches were grouped based on their primary focus area, as follows.

- Community-based behavior change approaches that create demand for sanitation and hygiene and change behavior:
  - CLTS,
  - School-Led Total Sanitation (SLTS),
  - PHAST,
  - Child Hygiene and Sanitation Training (CHAST), and
  - CHCs.
- Market-based approaches that develop or strengthen the market and supply chain for sanitation products and services.
- Financing approaches that use specific financing mechanisms to increase uptake or sustainability of sanitation among unserved or vulnerable populations.

In practice, some rural sanitation programs combine the approaches and cover more than one of the three focus areas. Understanding similarities and differences between the approaches can reveal opportunities and challenges for their integration or sequencing. Each approach aligns with certain characteristics, as discussed below.

---

[1] CLTS is the most widely used of these approaches.

[2] The main hygienic behaviors to target (particularly for children) are (i) handwashing with soap, (ii) personal hygiene, (iii) food hygiene, (iv) menstrual hygiene, and (v) safe household water management.

[3] PHAST is a "decision-support tool" developed by the World Bank and the World Health Organization to improve hygiene and sanitation behavior, reduce diarrheal disease, and encourage community management of water and sanitation services. It is a seven-step process of identifying and analyzing a problem, planning solutions, selecting options, constructing facilities, promoting behavior change, monitoring and evaluating activities, and conducting a participatory program evaluation.

Target populations (households, communities, and service providers) differ across the approaches:

- CLTS and SLTS target entire communities, school catchment areas, villages, or entire districts;
- PHAST, CHAST, and CHCs target subgroups within communities based on participation, club membership, or other targeting criteria (i.e., poverty mapping); and
- market-based approaches, microfinancing, and subsidy-based approaches target households.

Although there is a shared aim of improving rural sanitation, there are some differences between what is being monitored and used to assess success:[4]

- CLTS and SLTS aim to end open defecation within a geographic area;
- PHAST, CHAST, and CHCs are not exclusively focused on sanitation, and target additional improvements in hygiene, health, and nutrition-related behaviors; and
- market-based and microfinance approaches aim to increase latrine sales and repayment of loans.

The approaches use a variety of behavior change drivers:

- CLTS and SLTS rely on a mix of behavior change interventions, starting with emotional triggers (such as shock, disgust, shame, pride) along with health information and education, creating social norms and expectations to change defecation behaviors.
- PHAST, CHAST, and CHCs use rational health or hygiene messages and an educational approach.
- CHCs use health education through community clubs, reinforced with peer pressure and pride as emotional triggers.
- The market-based approaches use social marketing to expose latent demand for improved sanitation.

The approaches have different "philosophies" regarding their view of a participant in the intervention:

- PHAST, CHAST, and hardware subsidies view participants as beneficiaries needing assistance.
- CLTS, SLTS, and CHCs view participants as agents of community-level change, and at the same time beneficiaries of a behavior change intervention.
- Market-based and microfinance approaches view participants as (potential) customers willing to pay for a latrine, depending on the design, affordability, or other preferences.

A rural sanitation and hygiene program should incorporate the following principles:

- work with government, in coordination with sector stakeholders (including the health, education, finance, and environmental sectors);
- strengthen local systems area-wide, working across administrative units and targeting everyone (ensure inclusion);
- design the program based on the context, and evidence of what works in this context; and
- design the program to be flexible and adaptive, with constant learning about what works (and what does not).

---

4   All the approaches and their variations have a 100% household access target.

Appendix 3
COSTING RURAL SANITATION PROGRAMS

The costs of a rural sanitation and hygiene program need to be assessed and recorded to optimize the use of program resources, guide future programming, and enable comparisons of programs.[1] This requires cost tracking and reporting during all stages of the program, for which the following program activity groups can be used.

1. Planning
2. Formative Research
3. Program Mobilization
4. Capacity Development
5. Program Management
6. Community Implementation
7. Supply Strengthening
8. Sanitation Service Chain
9. Sanitation Finance
10. Monitoring and Evaluation
11. Sustainability Support
12. Environmental Sanitation
13. Enabling Environment
14. Knowledge Management

Activity-based categorization of costs helps ensure that all costs are identified, assessed, and reported. The costs should be reported by actors (who cover these costs) and by cost category. Actors may pay program expenditures directly, or make contributions of time, resources, etc. Common actors supporting a program are the following:

- program agency (main implementing agency; this is usually a government agency)
- government support (in addition to the program agency)
- external agencies
- private sector
- local entities
- community or households

The main steps in the costing process are as follows:

- During the planning stage:
  - select the main program components,
  - choose institutional arrangement(s) for the program implementation,
  - prepare an outline program budget, and
  - decide the costing approach to be applied during the program implementation.

---

[1] This section draws on the report, *Rural Sanitation Costing Guidance*, published in February 2019 and commissioned by WaterAid, Plan International, and United Nations Children's Fund (UNICEF).

- During the implementation stage:
  - track and report program costs.

- During the evaluation stage:
  - compile, analyze, and report on total program costs.

During the planning stage, the following also need to be reviewed and considered: (i) data sources; (ii) cost tracking tools; (iii) how to assess and present time contributions; and (iv) how to ensure that the program costs are comparable across programs, if required.

RECENT NATIONAL POLICY DOCUMENTS ON THE TREATMENT OF RURAL DOMESTIC WASTEWATER

In 2016, the State Council issued the 13th Five-Year Plan for Ecological Environmental Protection. This plan promotes a unified approach to the planning, construction, and management of rural wastewater treatment. It promotes wastewater and garbage treatment facilities and services, including the construction and renovation of toilets. The goal is to accomplish the treatment of at least 90% of rural domestic waste in administrative villages.

In 2016, the General Office of the Central Committee of the Communist Party of China and the General Office of the State Council issued the Opinions on the Full Implementation of the River Chief System. This focuses on the treatment of domestic wastewater and solid waste to improve the rural water environment and the living environment in villages.

In 2017, the Ministry of Environmental Protection and the Ministry of Finance jointly issued the 13th Five-Year Plan for the Comprehensive Improvement of the National Rural Environment. This plan emphasizes pollution control in villages with large populations and high densities to carry out rural domestic garbage and wastewater pollution control. It includes the construction of domestic wastewater treatment facilities, including collection networks (centralized treatment facilities or constructed wetlands, oxidation ponds, and other decentralized treatment facilities). The target is to reach a rural domestic wastewater treatment rate of more than 60%.

In 2018, the Central Committee of the Communist Party of China and the State Council issued the Opinions on the Implementation of the Rural Revitalization Strategy. This is a 3-year action plan for the improvement of rural human settlements, focused on treating rural garbage and wastewater, improving the appearance of villages, integrating resources, strengthening measures, and advancing governance in rural settlements. It includes promotion of the "toilet revolution" in rural areas, construction and upgrading of household sanitary toilets, and, at the same time, implementation of fecal pollution control, realizing full coverage of sanitary toilets in rural areas, and improving the quality of life of farmers. It emphasizes the strengthening of technical support and the adoption of wastewater treatment approaches that are most applicable to each region.

In 2018, the General Office of the Central Committee of the Communist Party of China and the General Office of the State Council issued the Three-Year Action Plan for the Improvement of Rural Human Settlement Environment. This action plan promotes combining pollution control and resource utilization, engineering measures and ecological measures, and concentration and decentralization, all adapted to local conditions. It also promotes the extension of urban wastewater networks to surrounding villages, as appropriate; low-cost, low-energy, easy-to-maintain, and high-efficiency wastewater treatment technologies; and encourages the use of ecological treatment processes. This is in addition to reducing the volume of domestic wastewater and recycling treated effluents. The action plan also addresses dredging of rivers, ponds, and ditches in front of houses to restore water ecology, gradually eliminating black and odorous water bodies, and bringing rural water environment governance into the river and lake chief systems of the PRC. The plan directed the provinces to formulate standards for rural domestic wastewater treatment and discharge, which all provinces have now done.

In 2018, multiple departments jointly issued the Guiding Opinions on Accelerating the Advancement of Agricultural Nonpoint Source Pollution Control in the Yangtze River Economic Zone. This document addresses the strengthening of the treatment of domestic wastewater in rural areas in a similar manner as the 3-year action plan mentioned above.

In 2018, the Ministry of Ecology and Environment and the Ministry of Agriculture and Rural Affairs issued the Action Plan for Pollution Control in Agriculture and Rural Areas. This action plan promotes the treatment of rural domestic wastewater in stages. All provinces should: distinguish drainage methods, discharge destinations, etc.; speed up the formulation and revision of treatment and discharge standards; and select practical technologies and models suitable for the region. It emphasizes the importance of ensuring the long-term operation of the pollution control facilities.

In 2018, the Ministry of Ecology and Environment and the Ministry of Housing and Urban–Rural Development issued the Notice on Accelerating the Establishment of Local Rural Domestic Wastewater Discharge Standards. Formulation of discharge standards for rural domestic wastewater treatment should be based on local conditions, the degree of population density in a village, the scale of wastewater generation, discharge direction, and the improvement needs of the local environment, while prioritizing reuse and keeping a focus on results and ease of supervision.

In 2019, the Ministry of Housing and Urban–Rural Development issued the Technical Standards for Rural Domestic Wastewater Treatment Facilities (GB/T51347-2019). Based on this standard, rural domestic wastewater treatment should adopt the biofilm method (anaerobic biofilm method, biological contact oxidation tank, biological filter, biological turntable, etc.), activated sludge method (activated sludge method, oxidation ditch activated sludge method, membrane Bioreactors, etc.), natural biological treatments (constructed wetlands, stable ponds, etc.), and physical and chemical methods (grids, grit tanks, regulating tanks, and chemical dephosphorization, etc.). The sludge treatment can be natural drying, composting, or co-processing with rural solid organic matter, or the sludge may enter a municipal system to be treated together with municipal sludge. Prefabricated devices can be used for household processing. The residence time of wastewater in the septic tank should be 24–36 hours.

In 2019, the Ministry of Ecology and Environment, in conjunction with the Ministry of Agriculture and Rural Affairs, compiled the Guidelines for the Compilation of Water Pollutant Discharge Control Specifications for Rural Domestic Wastewater Treatment Facilities (Trial). These guidelines provide clarification of the 2018 Notice on Accelerating the Establishment of Local Rural Domestic Wastewater Treatment Discharge Standards regarding the scope of application, classification and grading, determination of control indicators, and control requirements.

In 2019, nine agencies including the Central Agricultural Office, the Ministry of Agriculture and Rural Affairs, and the Ministry of Ecology and Environment jointly issued the Guiding Opinions on Promoting the Treatment of Rural Wastewater. This document emphasizes that rural wastewater treatment should: be based on the PRC's actual situation; take wastewater reduction, classified on-site treatment, and recycling as a guide; strengthen overall planning; and clarify the overall requirements and key tasks by region. The key tasks include getting a comprehensive view of the current situation, formulating action plans scientifically, selecting technical models rationally, speeding up the formulation and revision of standards, and improving construction and management mechanisms.

In 2020, the Central Committee of the Communist Party of China and the State Council issued the Opinions on Promoting in the Field of "Three Rurals" to Ensure the Realization of a Well-Off Society in an All-Round Way as Scheduled. This document promotes rural wastewater treatment and gives priority to townships and central villages, recommends treatment of black and odorous water bodies, supports farmers to carry out village cleaning and greening, promotes the construction of "beautiful homes," and encourages qualified localities to subsidize the maintenance of public facilities in rural areas.

In 2021, the Central Committee of the Communist Party of China and the State Council issued the Opinions on Comprehensively Promoting Rural Revitalization and Accelerating Agricultural and Rural Modernization. This includes the implementation of a 5-year action plan to improve the rural living environment. It encourages the coordination of toilet improvements and wastewater, black, and odorous water treatment in rural areas, building wastewater treatment facilities according to local conditions, and improving the management and protection mechanism of rural facilities.

In 2021, the General Office of the Ministry of Agriculture and Rural Affairs and the General Department of the National Rural Development Bureau issued the Guidelines for Social Capital Investment in Agriculture and Rural Areas. This encourages social capital to participate in the above 5-year action plan, including in the construction and operation of initiatives such as the rural toilet revolution, rural wastewater treatment, and rural domestic sewage treatment. Public welfare projects for stable income in agricultural and rural areas should be actively explored, and the public–private partnership model should be promoted.

In 2021, the General Office of the Central Committee of the Communist Party of China and the General Office of the State Council issued the Five-Year Action Plan for the Improvement of Rural Human Settlements (2021–2025). This plan prioritizes the Beijing–Tianjin–Hebei Area, the Yangtze River Economic Belt, the Guangdong–Hong Kong–Macao Greater Bay Area, the Yellow River Basin, and control unit areas where water quality needs improvement. It focuses on the improvement of water source protection areas and urban–rural fringes, township government residences, central villages, and tourist scenic spots. It encourages pilot projects for rural wastewater treatment in areas such as plains, mountains, hills, and water shortage areas, as well as in sensitive ecological environments. Guided by resource utilization and sustainable treatment, rural wastewater treatment technologies should meet the local conditions. It emphasizes giving priority to treatment technologies with low operating costs and simple maintenance.

RECENT PROVINCIAL POLICY DOCUMENTS ON RURAL WASTEWATER TREATMENT

Province	Release	Policy Document Title
Shanghai	2018	Shanghai Rural Domestic Wastewater Treatment Facilities Operation and Maintenance Management Measures (Trial)
Beijing	2019	Three-Year Action Plan for Urban and Rural Water Environment Governance in Beijing (July 2019–June 2022)
Chongqing	2019	Work Plan for Rural Domestic Wastewater Treatment in 2019 in Chongqing
Guizhou	2019	Special Action Plan for Rural Domestic Wastewater Treatment in Guizhou Province (2019–2020)
Hainan	2019	Measures for the Assessment of Rural Domestic Wastewater Treatment in Hainan Province
Henan	2019	Opinions on Accelerating the Promotion of Rural Domestic Wastewater Treatment
Shandong	2019	Rural Domestic Wastewater Treatment Action Plan in Shandong Province
Fujian	2020	2020 Implementation Plan for Rural Domestic Wastewater Treatment in Fujian Province
Guangxi	2020	Rural Domestic Wastewater Treatment Implementation Plan in Guangxi
Hebei	2020	Work Plan for Rural Domestic Wastewater Treatment in Hebei Province (2021–2025)
Heilongjiang	2020	County Rural Domestic Wastewater Treatment Special Plan in Heilongjiang
Hunan	2020	Guiding Opinions on Special Planning for Rural Domestic Wastewater Treatment in Hunan Province Technical Guidelines for Rural Domestic Wastewater Treatment in Hunan Province (Trial)
Jilin	2020	Action Plan for Promoting Rural Domestic Wastewater Treatment in Jilin
Shanxi	2020	Measures for the Operation and Management of Rural Domestic Wastewater Treatment Facilities in Shanxi (for Trial Implementation) Guide to Rural Domestic Wastewater Treatment in Shanxi Province
Sichuan	2020	Three-Year Promotion Plan for Rural Domestic Wastewater Treatment in Sichuan Province
Tianjin	2020	Administrative Measures for the Operation and Maintenance of Rural Domestic Wastewater Treatment Facilities in Tianjin

Source: Compilation by author as presented in Asian Development Bank. 2022. *Rural Wastewater Management in the People's Republic of China*. Consultant's report. Manila (TA 9825-PRC).